COMPLETING THE "BIG DIG"

Managing the Final Stages of Boston's Central Artery/Tunnel Project

Committee for Review of the Project Management Practices Employed on the Boston Central Artery/Tunnel ("Big Dig") Project
Board on Infrastructure and the Constructed Environment
Division on Engineering and Physical Sciences
NATIONAL ACADEMY OF ENGINEERING
NATIONAL RESEARCH COUNCIL
TRANSPORTATION RESEARCH BOARD *OF THE NATIONAL ACADEMIES*

THE NATIONAL ACADEMIES PRESS
Washington, D.C.**www.nap.edu**

THE NATIONAL ACADEMIES PRESS500 Fifth Street, N.W.Washington, DC20001

NOTICE: The project that is the subject of this report was approved by the Governing Board of the National Research Council, whose members are drawn from the councils of the National Academy of Sciences, the National Academy of Engineering, and the Institute of Medicine. The members of the committee responsible for the report were chosen for their special competences and with regard for appropriate balance. This study was supported by Contract No. 02-1218 between the National Academy of Sciences and the Massachusetts Turnpike Authority. Any opinions, findings, conclusions, or recommendations expressed in this publication are those of the author(s) and do not necessarily reflect the views of the organizations or agencies that provided support for the project.

International Standard Book Number 0-309-08887-9

Available from:
Board on Infrastructure and the Constructed Environment
National Research Council
500 5th Street, N.W.
Washington, DC 20001
Printed in the United States of America

THE NATIONAL ACADEMIES

National Academy of Sciences
National Academy of Engineering
Institute of Medicine
National Research Council

The **National Academy of Sciences** is a private, nonprofit, self-perpetuating society of distinguished scholars engaged in scientific and engineering research, dedicated to the furtherance of science and technology and to their use for the general welfare. Upon the authority of the charter granted to it by the Congress in 1863, the Acade my has a mandate that requires it to advise the federal government on scientific and technical matters. Dr. Bruce M. Alberts is president of the National Academy of Sciences.

The **National Academy of Engineering** was established in 1964, under the charter of the National Academy of Sciences, as a parallel organization of outstanding engineers. It is autonomous in its administration and in the selection of its members, sharing with the National Academy of Sciences the responsibility for advising the federal government. The National Academy of Engineering also sponsors engineering programs aimed at meeting national needs, encourages education and research, and recognizes the superior achievements of engineers. Dr. Wm. A. Wulf is president of the National Academy of Engineering.

The **Institute of Medicine** was established in 1970 by the National Academy of Sciences to secure the services of eminent members of appropriate professions in the examination of policy matters pertaining to the health of the public. The Institute acts under the responsibility given to the National Academy of Sciences by its congressional charter to be an adviser to the federal government and, upon its own initiative, to identify issues of medical care, research, and education. Dr. Harvey V. Fineberg is president of the Institute of Medicine.

The **National Research Council** was organized by the National Academy of Sciences in 1916 to associate the broad community of science and technology with the Academy's purposes of furthering knowledge and advising the federal government. Functioning in accordance with general policies determined by the Academy, the Council has become the principal operating agency of both the National Academy of Sciences and the National Academy of Engineering in providing services to the government, the public, and the scientific and engineering communities. The Council is administered jointly by both Academies and the Institute of Medicine. Dr. Bruce M. Alberts and Dr. Wm. A. Wulf are chair and vice chair, respectively, of the National Research Council.

www.national-academies.org

COMMITTEE FOR REVIEW OF THE PROJECT MANAGEMENT PRACTICES EMPLOYED ON THE BOSTON CENTRAL ARTERY/TUNNEL ("BIG DIG") PROJECT

JOHN T.CHRISTIAN, *Chair,* Consulting Engineer, Waban, Massachusetts
JAMES G.BROWDER, JR., Wilbur Smith Associates, Richmond, Virginia
JOHN C.DAVIS, General Motors Corporation (retired), Oxford, Michigan
CHRIS T.HENDRICKSON, Carnegie Mellon University, Pittsburgh, Pennsylvania
JAMES LEFTER, Virginia Tech, Crofton, Maryland
RICHARD K.SANDAAS, Municipality of Metropolitan Seattle (METRO) (retired), Kirkland, Washington
RICHARD P.WEAVER, California DOT (retired), Big Arm, Montana

Staff

RICHARD G.LITTLE, Director, Board on Infrastructure and the Constructed Environment
MICHAEL COHN, Program Officer
JASON DREISBACH, Research Associate
DANA CAINES, Financial Associate
PAT WILLIAMS, Senior Project Assistant

Acknowledgment of Reviewers

This report has been reviewed in draft form by individuals chosen for their diverse perspectives and technical expertise, in accordance with procedures approved by the NRC's Report Review Committee. The purpose of this independent review is to provide candid and critical comments that will assist the institution in making its published report as sound as possible and to ensure that the report meets institutional standards for objectivity, evidence, and responsiveness to the study charge. The review comments and draft manuscript remain confidential to protect the integrity of the deliberative process. We wish to thank the following individuals for their review of this report:

Frances T.Banerjee, Banerjee and Associates

William W.Badger, Arizona State University

Jack E.Buffington (NAE), University of Arkansas

Robert A.Frosch (NAE), Harvard University

Delon Hampton (NAE), Delon Hampton and Associates

David Luberoff, Harvard University

Sue McNeil, University of Illinois

Charles L.Miller, Baker Transportation

Norman A.Nadel (NAE), Nadel Associates

Kenneth N.Reinschmidt (NAE), Texas A&M University

Although the reviewers listed above have provided many constructive comments and suggestions, they were not asked to endorse the conclusions or recommendations, nor did they see the final draft of the report before its release. The review of this report was overseen by Morris Tanenbaum (NAE), AT&T Corporation (retired), and Lloyd Duscha (NAE), U.S. Army Corps of Engineers (retired). Appointed by the National Research Council, they were responsible for making certain that an independent examination of this report was carried out in accordance with institutional procedures and that all review comments were carefully considered. Responsibility for the final content of this report rests entirely with the authoring committee and the institution.

Contents

List of Figures

Acronyms and Abbreviations

ACEC/MA	American Council of Engineering Companies of Massachusetts
B/PB	Bechtel/Parsons Brinckerhoff
BCCF	Budget, Cost, and Commitment Forecast
BCI	Building Cost Index
BICE	Board on Infrastructure and the Constructed Environment
CA/T	Central Artery/Tunnel
CCI	Construction Cost Index
CEVP	Cost Estimating Validation Process
CIM	Construction Industries of Massachusetts
CIS	Construction Information System
CSU	Cost and Schedule Update
DOT	Department of Transportation
DRB	Dispute Review Board
EIS	Environmental Impact Statement
FHWA	Federal Highway Administration
GAO	Government Accounting Office
ICE	Interstate Cost Estimate
IPO	Integrated Project Organization
MDPW	Massachusetts Department of Public Works
MHD	Massachusetts Highway Department
MTA	Massachusetts Turnpike Authority
MTA CEO	Massachusetts Turnpike Authority Chief Executive Officer
NAE	National Academy of Engineering
NRC	National Research Council
OCIP	Owner Controlled Insurance Program
PMM	Project Management Monthly
RFI	Request for Information
TRB	Transportation Research Board
VECP	Value Engineering Change Proposal

Executive Summary

The Central Artery/Tunnel (CA/T) project is one of the most complex, and certainly the most costly, public-infrastructure projects ever undertaken in the United States. Often called the "Big Dig," the project was conceived to improve the traffic flow in downtown Boston and link several major roadways and transportation hubs. It was intended to replace a badly deteriorated and congested elevated roadway (I-93, the "Central Artery" through Boston), extend the Massachusetts Turnpike (I-90) to Logan Airport through a harbor tunnel, provide an interchange of I-90 and I-93, and replace the I-93 bridge over the Charles River. The project was originally estimated to cost $2.6 billion (1982 dollars) and be completed in 1998. The current estimate is $14.6 billion (2002 dollars) with completion anticipated in 2005. All design work has essentially been done, construction is about 85 percent complete, and portions of the project are now being put into operation.

The overall responsibility for the CA/T project was initially assigned to the Massachusetts Department of Public Works, which became the Massachusetts Highway Department (MHD) in 1991. The MHD hired a joint venture of the firms Bechtel and Parsons Brinckerhoff Quade and Douglas, hereinafter referred to as B/PB, to serve as project manager. In 1997, the state legislature designated the Massachusetts Turnpike Authority (MTA) to be owner/operator of the overall Metropolitan Highway System, with consequent responsibility for management of the CA/T project. That management function is now performed by an integrated project organization (IPO) of MTA staff and B/PB staff who essentially work as one single organization under the direction of the MTA.

The Central Artery/Tunnel project has had many noteworthy technological accomplishments that will serve to guide future urban construction. Among them are the deep slurry walls constructed in soft clay, the soil-freezing and tunnel-jacking operation at the Fort Point Channel crossing, and the Leonard P.Zakim Bunker Hill Bridge, which is the widest cable-stayed bridge in the world. It must also be noted that extensive construction has taken place in a dense urban area with minimal damage to existing structures and utilities, traffic has been kept flowing through a busy city for over a decade, and a major railroad yard continued operations while a tunnel was built beneath it.

Accomplishments have not been limited to the technological. The CA/T project team has been sensitive to the needs and desires of the communities that were inevitably affected by the construction activities. The Owner-Controlled Insurance Program (OCIP) was an effective and cost-reducing response to the challenge of obtaining adequate insurance coverage for the large numbers of engineering and contracting firms involved in the project. And the project's safety record for 2002–5.5 recordable worker injuries per 100 full-time employees—is significantly below the national average of 8.2.

This report was requested by the MTA, which sought an independent assessment of the CA/T project's current project-management and contract-administration practices. The MTA wished to determine if such practices will be adequate for completing the project in a timely and cost-effective manner. The report therefore focuses on project-management and contract-administration practices and procedures in place as of October 2002 and looks toward the future—that is, to the end of the project and its transition to a fully operational system. The report does not attempt to address the many past controversies that have surrounded the project's justification, funding, cost, schedule, or management; it considers prior project performance, cost and schedule escalation, and financing plans only to the extent that they may affect future performance.

The assessment was performed by an expert committee appointed by the National Research Council (NRC). The NRC's Board on Infrastructure and the Constructed Environment (BICE), in collaboration with the National Academy of Engineering (NAE) and the NRC's Transportation Research Board (TRB), helped to oversee the study process. The committee members who undertook this study had expertise in the key areas of project management, contract administration and management, and transportation-agency administration, but they had no significant relationship with the project, the MTA, or the management consultant B/PB. The committee was constrained to accept the information presented by the MTA and IPO and information included in prior audit reports.

The committee was charged with the following five tasks:

1. Determine the adequacy of current procedures for addressing the CA/T project's cost and schedule issues.
2. Recommend refinements to the organization and management of its contract-administration function.
3. Recommend improvements to the reporting and controls procedures.
4. Determine the adequacy of oversight of the project's management consultant.
5. Determine the appropriateness and effectiveness of the structure for transition from construction to operations.

In undertaking these tasks, the committee was well aware of the many controversies that have surrounded the CA/T project. During its long history, the project has been scrutinized by myriad investigative and oversight bodies, and it will no doubt be subject to additional probing yet to come. However, the committee went forward with the rationale that there is $2 billion still to be spent, major construction to complete, and a key element of the Massachusetts transportation system to be made fully operational. The committee maintains that management and organizational changes can still be implemented both for saving money and easing, even expediting, the transition to operations. The findings and recommendations presented below are thus offered in the belief that the future of the CA/T project is still to be determined and that every opportunity for efficiency and improvement should be evaluated and acted on.

The committee's findings and recommendations are summarized below under headings that correspond to its five tasks. The committee considers all of its conclusions to be relevant to the issues confronting the MTA, but it also believes that two matters of overarching importance merit special emphasis and priority attention:

- The uncertain financial burden and potential for delay posed by the excessive number and monetary value of outstanding contract claims and related issues.
- The need for the transition process from construction to operation and maintenance to be properly organized and managed.

FINDINGS AND RECOMMENDATIONS

Cost and Schedule Procedures

Schedule Management

Finding 1: Meeting the CA/T's project's scheduled targets continues to be a problem. Despite an emphasis on reaching the milestones on time, slippage continues, thereby reducing public confidence in CA/T management.

Recommendation 1: CA/T project managers should strategically evaluate future schedules by determining what critical tasks need to be done without fail and how long these activities will likely take. Published completion dates should be developed around realistic workflows and schedule risks, with modest allowances for unknown issues.

Cost Management

Finding 2: The project has had large cost increases resulting from changes in scope, design, and project limits, as well as from deficiencies coordinating contracts.

Recommendation 2: All contract modifications should be comprehensively reviewed—prior to execution—for impacts on scope, design details, interfaces, and contract duration.

Cost/Schedule Tradeoff

Finding 3: The committee could not identify sufficient analysis to justify all construction-acceleration efforts that have occurred.

Recommendation 3: The MTA should rigorously analyze all costs associated with project acceleration, especially for completing the key "critical path" tasks (those in which a delay could cause critical project milestones to be missed).

Value Engineering

Finding 4: The savings from contractor-proposed "value engineering change proposals" (VECPs) to date has been $25 million, which is a negligible amount compared with the constructed value of the project.

Recommendation 4: The MTA should aggressively promote the VECP for the remaining work. This effort should increase contractor level of awareness, encourage innovation, and yield cost reductions.

Contract Administration

Claims Management

Finding 5a: The project's exposure to unsettled claims and changes is significant. Both the number of claims and their average age have been growing steadily for the past 4 years.

Finding 5b: Responsibility for resolving claims and changes for amounts over $250,000 is currently assigned to the legal staff of the MTA, and the imminent departure of a key staff member may adversely affect staff productivity.

Recommendation 5a: Resolution of unresolved changes and claims should be deemed a high-priority issue, meriting immediate attention by the MTA. It constitutes a project within the project, to which project management principles should be applied.

Recommendation 5b: A closeout schedule should be developed, along with identification of resources necessary to support it. A target date of July 2004 should be established for resolution of the unsettled claims and changes.

Recommendation 5c: Consideration should be given to paying contractors for the estimated direct costs of directed changes and recognized claims in advance of settling other entitlement issues. In that way, the burdens on contractors and subcontractors carrying these costs can be relieved.

Contingency Management

Finding 6: While individual contracts have been analyzed for exposure to changes, comprehensive risk- and contingency-management tools and processes do not appear to be in place.

Recommendation 6: The MTA should initiate contingency-management procedures that include continual comprehensive risk analysis to quantify and refine contract contingencies, individual-contract contingency tracking, and a contingency drawdown plan that includes contingency-use forecasts.

Cost Recovery

Finding 7: The current cost-recovery program emphasizes legal issues prior to resolution of engineering issues. This process is reported to have replaced a partnering-style relationship with an adversarial process.

Recommendation 7: The MTA should take steps to reinstate the engineering-driven process in order to resolve cost-recovery issues.

Reporting and Controls

Senior Management's Information Overload

Finding 8: Too much uncoordinated information is provided to the project's senior management, and project reports do not lend themselves to ready analysis—particularly with regard to such concerns as the uncertainty of forecasts and the trade-off possibilities between time reductions and cost increases.

Recommendation 8: As the project foci shift—to managing claims, cost-recovery processes, and the transition to operation—project reports should also change to emphasize summary information and strategic issues.

Schedule Analysis

Finding 9: The MTA has developed an aggressive and optimistic schedule for project completion. However, the project milestones are prone to slippage as events occur.

Recommendations 9: Project reports should reflect the probability of meeting milestones or the probabilistic distribution of completion times, based on the results of past efforts.

Oversight for Management Consultant

Integrated Project-Management Oversight

Finding 10: The integrated project organization (IPO) structure currently utilized by the MTA to direct the design and construction of the CA/T project appears to be functioning reasonably well. Although most of the contracts are now under way or completed, the committee observed that a detailed plan for downsizing and ultimately eliminating the management-consultant staff was not in place.

Recommendation 10: MTA should implement an aggressive plan to downsize the B/PB staff members who are not essential to completing contracts and resolving claims. At the same time, the MTA should ensure that key staff members remain in place to finalize and close out all contracts and claims.

Finding 11: The existing management tools and metrics currently in use are sufficient for overseeing the management consultant.

Recommendation 11: If the MTA wishes to improve its oversight of the management consultant, the committee encourages it to make better use of the tools presently available rather than develop new tools.

Independent Peer Review

Finding 12: The CA/T project has had numerous reviews over time for different purposes. However, there has not been a continuing independent peer-review of technical issues, management, and strategic directions.

Recommendation 12: The MTA should establish an external, independent peer-review program to address technical and management issues until transition to operations and maintenance is completed. The frequency of reviews and peer-review participants should vary in accordance with the issues to be addressed.

Transition from Construction to Operations

Public Education

Finding 13: Information transfer to the motoring public has been minimal thus far and not well timed for the scheduled opening of the first completed sections of the project.

Recommendation 13: MTA should have a media team, with technical assistance provided by the project staff, to develop and immediately implement an innovative plan for educating the public to use the CA/T.

Strategic Thinking

Finding 14: The committee has not seen an adequate plan for guiding the transition from a construction organization dominated by project-management consultants to an operations organization that is largely composed of full-time MTA staff.

Recommendation 14: The MTA should establish processes to assist its chairman in identifying and prioritizing strategic issues and in carrying out the transition from design and construction to operations and maintenance.

Security

Finding 15: During its October site visit, the committee observed what it perceived as a lack of security.

Recommendation 15: A comprehensive security program should be developed, immediately implemented, and maintained not only until completion of the project but also throughout its transition to operation.

1

Introduction

BACKGROUND

This report provides an assessment of project-management and contract-administration practices currently being employed for the Central Artery/Tunnel (CA/T) Project, often called the "Big Dig," in Boston, Massachusetts. The project was originally estimated to cost $2.6 billion (base year 1982 dollars, according to the Final Environmental Impact Statement (EIS) approved by the Federal Highway Administration in 1985) and was to be completed in 1998. Now it is estimated cost is $14.6 billion in current dollars ($8.0 billion in 1982 dollars) and completion is expected in 2005. The design of the project has essentially been done, all major contracts have been awarded, and construction is approximately 85 percent complete. The CA/T project consists of three major parts. The first is a highway tunnel that replaces a 50-year-old badly deteriorated and congested elevated highway known as the Central Artery (I-93) that traverses downtown Boston from North to South. The original highway had been designed to carry 75,000 vehicles daily but is now used by 200,000. Removal of the elevated highway will create 27 acres of open space. The second part of the project includes extension of the Massachusetts Turnpike (I-90) to Logan Airport through a third harbor crossing, known as the Ted Williams Tunnel. This segment will also allow northbound traffic on I-93 to interchange with east-west traffic on I-90 and connect with the Ted Williams Tunnel to Logan Airport. The interchange requires a series of tunnels, viaducts, and ramps operating at five different levels. The third part of the project is a bridge and its associated interchange system, which will replace the I-93 bridge over the Charles River. In summary, the CA/T project's new tunnels and bridges are intended to alleviate serious traffic congestion, eliminate a troublesome eyesore, reconnect old neighborhoods, and create open space in the middle of a historic city. The project map is shown in Figure 1.1.

ENGINEERING CHALLENGES

New construction consists of 161 lane miles, almost half of them underground, in 7.5 miles of right-of-way through downtown Boston and the surrounding areas and many new on/off ramps tied into the existing surface streets. In the process of construction, underground utilities relating to gas, water, electric, sewer, and telephone, many of them dating from the 19th century, have been relocated and modernized. All of this will have been accomplished while keeping the downtown Boston area fully operational. Engineering accomplishments have included erection of the widest cable-stayed bridge in the world, the biggest use of slurry wall construction in North America, jack tunneling under active railroad tracks through unstable soils, and other novel technologies for highway bridges and submersible tunnels.

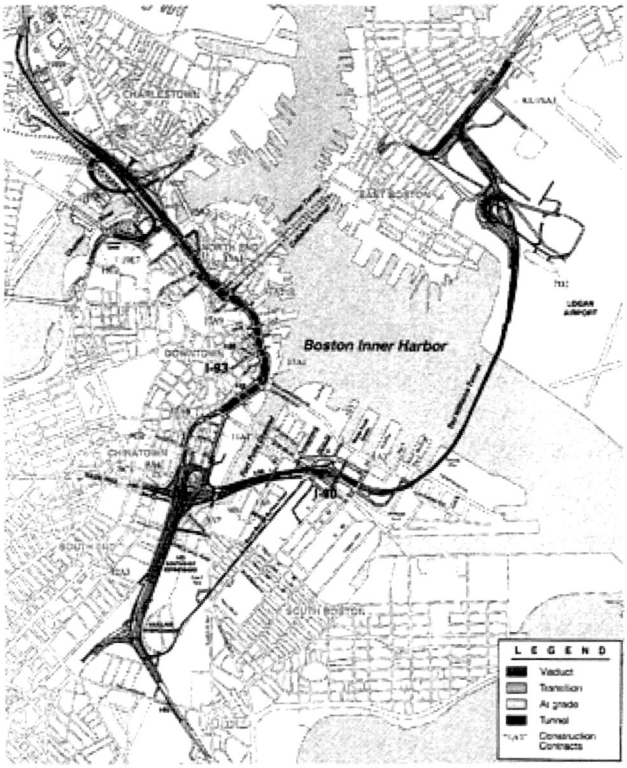

FIGURE 1.1 Project Map (Source: http://www.bigdig.com/thtml/maps01.htm)

The project will be opened in stages: the Ted Williams Tunnel, already used by weekday commercial traffic for several years, was opened to all weekday traffic in January 2003 in conjunction with the extension of I-90; the northbound lanes of the new Central Artery are scheduled to be opened in March 2003, and the southbound lanes in February 2004. The old Central Artery is scheduled to be demolished later in 2004.

The CA/T project conducted extensive outreach to involve stakeholders during the project's planning and design, which resulted in many modifications and extensive mitigation measures. For just the air rights component alone (pertaining to development projects above roadways), over 70 meetings were held between 1998 and 2000, and a master plan, called "Civic Vision for Turnpike Air Rights in Boston," was adopted by the MTA and the City of Boston. The existence of such active citizen involvement and consensus-building by state and local agencies suggest that stakeholders are likely to continue their involvement in the project up to its completion.

PROJECT MANAGEMENT

The state's Secretary of Transportation initially assigned overall responsibility for the CA/T project to the Massachusetts Department of Public Works (MDPW), which became the Massachusetts Highway Department (MHD) in 1991. This assignment carried a proviso that the agency hire an exceptionally experienced and capable contractor as project manager. In 1985, MDPW entered into an agreement with a joint venture of the firms Bechtel and Parsons Brinckerhoff Quade & Douglas, hereinafter referred to as B/PB in this report, to prepare a preliminary project management plan. Later that year, MDPW issued a 1-year contract to B/PB to develop a comprehensive work plan for managing and reviewing design consultants. This contract was followed by a number of limited-term contracts, called "work programs," extending to the present. Ultimately, B/PB coordinated and managed all project activities by both representing and advising the owner (the Commonwealth of Massachusetts). The owner retained the authority to make actual decisions, but B/PB had substantial control over design and construction.

In 1997, the state legislature designated the Massachusetts Turnpike Authority (MTA) the owner/operator of the Metropolitan Highway System with responsibility for management of the CA/T project. Funding to this day, however, still goes through MHD. Nevertheless, the CA/T project team reports that the current arrangement, whereby the Federal Highway Administration (FHWA), MHD, and MTA coordinate the funding and oversight of the project, is working efficiently.

Currently, the overall management of the CA/T project is performed by an integrated project organization (IPO) composed of MTA staff and B/PB staff mixed in at various levels into one single organization. The project director is an MTA employee. Further discussion of the project's organizational structure can be found throughout the remainder of this report.

EXTERNAL OVERSIGHT

A number of major independent reviewers have analyzed the CA/T project's management and organizational structure, cost savings, effectiveness and efficiency, and change-order process. First, the FHWA has responsibility for oversight of the project as a whole because it distributes federal funds via the U.S. Department of Transportation (DOT). FHWA is responsible for ensuring that the project adheres to federal regulations and standards, and it is expected to evaluate the state's programs critically and provide technical assistance as necessary. Many of the technical aspects of the project's contracts, including changes, are approved by FHWA. The DOT Inspector General and independent review panels have in turn reviewed the FHWA oversight activities. State agencies, including the Massachusetts State Auditor and Inspector General, provide reviews of various costs, revenues, and processes. The Massachusetts State Office of Administration and Finance contracts with a major auditing firm to provide a DOT-mandated evaluation of the yearly Cost and Schedule Update (CSU). The U.S. General Accounting Office (GAO) has also performed various audits of the CA/T project. Additionally, bondholders, the press, and local citizens' groups, including the Artery Business Committee and the Massachusetts Taxpayers Foundation, follow the CA/T project closely.

NATIONAL RESEARCH COUNCIL INVOLVEMENT

The present independent assessment of project-management and contract-administration practices now being employed for the CA/T project was requested by the MTA; the assessment's aim was to determine whether such practices are adequate to bring the project to completion in a timely and cost-effective manner. That assessment, resulting in this report, reviewed procedures in place as of October 2002 and only considered prior project performance, cost and schedule escalation, and financing plans only to the extent that they may affect future performance.

The assessment was performed by an expert committee appointed by the National Research Council (NRC). The NRC's Board on Infrastructure and the Constructed Environment (BICE), in collaboration with the National Academy of Engineering (NAE) and the NRC's Transportation Research Board (TRB), helped to oversee the study process. Experts were screened to ensure no recent direct involvement on their part with the project or its management consultant. The committee members who undertook this study had expertise in the key areas of project management, contract administration and management, and transportation-agency administration. Biographies of committee members are presented in Appendix A.

The committee's task was to address the following five specific issues:

1. The adequacy of current procedures for balancing the CA/T project's cost and schedule issues and the degree to which its multiplicity of projects and priorities are being successfully organized and implemented.
2. Refinements to the organization and management processes for the contract-administration function, including the control of change orders, the contractor-claims process, the execution of contract modifications, and the review of cost-recovery issues.
3. Recommended improvements to the reporting and controls procedures, including determination of whether the appropriate amount of information is being provided to project management for decision-making purposes.
4. Adequacy of methods employed by the owner (MTA) to oversee the management consultant (B/PB).
5. The appropriateness and effectiveness of the structure for the project's transition from construction to an operable transportation-infrastructure system.

This report addresses each of these issues separately in the five chapters that follow. Its intention is not to identify causes or assign responsibility for past project performance but to provide an assessment of the present situation and its potential future. However, the report does include a brief review of relevant project history in order to establish context for the findings and recommendations.

The committee could not conduct an independent audit in the time allotted for this study and was therefore constrained to accept the information included in prior audit reports and what was directly presented to it by the MTA and IPO. The committee did not review design or construction details of the individual projects.

MEETINGS AND BRIEFINGS

Most of the information analyzed in this study was obtained from the CA/T project staff. The committee met in Boston for two 3-day meetings in October and November of 2002. During those meetings, the CA/T project staff—augmented by representatives from B/PB, MTA, FHWA, and other agencies—briefed the committee on aspects of the project relating to the five specific issues it was charged to address. The committee also toured the entire site of the project, with project staff as guides, during one October afternoon. In addition, a one-day roundtable discussion was held in Washington, D.C., in December 2002 to allow the committee to hear from both contractors and engineering consultants on the CA/T project. Among the participants in the discussion were representatives from Construction Industries of Massachusetts (CIM), the American Council of Engineering Companies of Massachusetts (ACEC/MA), and the CA/T project management team.

REFERENCES

Massachusetts Turnpike Authority (MTA). 2002. "Maps+Plans." Available online at http://www.bigdig.com/thtml/maps01.htm [December 31, 2002].

2

Cost and Schedule Procedures

INTRODUCTION

The focus of this report is the procedures that are currently in place (October 2002) to control costs and schedules of the project through its projected completion in 2005. However, when trying to envision a project's future it is important to be aware of its past, and the past performance of the CA/T project has been notorious for cost growth and schedule delay.

BACKGROUND

Since the project's Final EIS approval by FHWA in 1985, costs (in constant dollars) have grown to more than three times the original estimate (Figure 2.2) and the duration has increased by 6 years (Figure 2.3). Analysis of the project's performance presented to the committee by the CA/T project management team indicated that about half of the cost growth was caused by inflation[1] (the original estimates were in 1982 dollars, as required by FHWA) and that a portion of this could be attributed to the extended schedule. The analysis also indicated that 3 percent of the increased cost resulted from efforts to maintain the schedule and about 28 percent derived from increased scope and unplanned expenses to mitigate the project's effect on the environment and the movement of traffic in Boston (Figure 2.1). The project-management team asserted that the increases would have been greater if they had not initiated procedures to control the growth of costs and schedule. For example, the project team claims to have saved $750 million through approved cost-containment initiatives, as well as $500 million saved by deploying its Owner Controlled Insurance Program (OCIP).

The cost and schedule performance problems may be attributed to many different sources, including the low original estimate, inflation, cost of mitigation, unexpected technical complexities, delays in making decisions, and changes during construction. A recent review of large public works projects over the last century concluded that they are consistently underestimated, a phenomenon attributed to the desire of the project advocates to have their projects approved (Flyvbjerg, 2002). Others have reached similar conclusions (Wachs, 1989; Merrow, 1988), as did a writer in the local press, who characterized the project's original cost estimate as follows:

[1]The "absolute" cost growth for this project, without considering the change in the value of money over time, is approximately $12.0 billion (current project cost estimate of $14.6 billion minus original project cost estimate of $2.6 billion). The project management team asserts that about half (approximately $6.5 billion) of that $12.0 billion can be attributed to inflation from 1985, when expenditures on the project began. The estimate of the effect of inflation is derived from the *Engineering News-Record's* Building Cost Index (BCI) and Construction Cost Index (CCI) combined into a single index. An escalation because of inflation is calculated for each year of the project by applying the index to the actual or projected annual expenditures, thereby determining their value in 1982 dollars. These yearly escalations (actual or projected expenditures minus their 1982 value) were summed up to arrive at the total escalation of approximately $6.5 billion.

> [The Big Dig sponsors'] early estimates were deliberately optimistic. Politically, they had no choice. If they'd been honest, the project never would have happened, because Americans are unwilling to face the actual cost of great civic projects (Campbell, 2000).

The original estimate was developed in 1982 dollars, without allowance for inflation or mitigation, and was presented in the 1985 Environmental Impact Statement (EIS) before detailed technical studies had been undertaken. The 1982 estimate was based on guidelines in FHWA's Interstate Cost Estimate (ICE) manual, which excluded inflation. It was thought that inflation would be captured during biennial updates. Inflation has indeed had a significant impact on the project cost, but the impact is difficult to assess accurately because of the dynamic interplay of the project's scope, schedule, and construction costs.

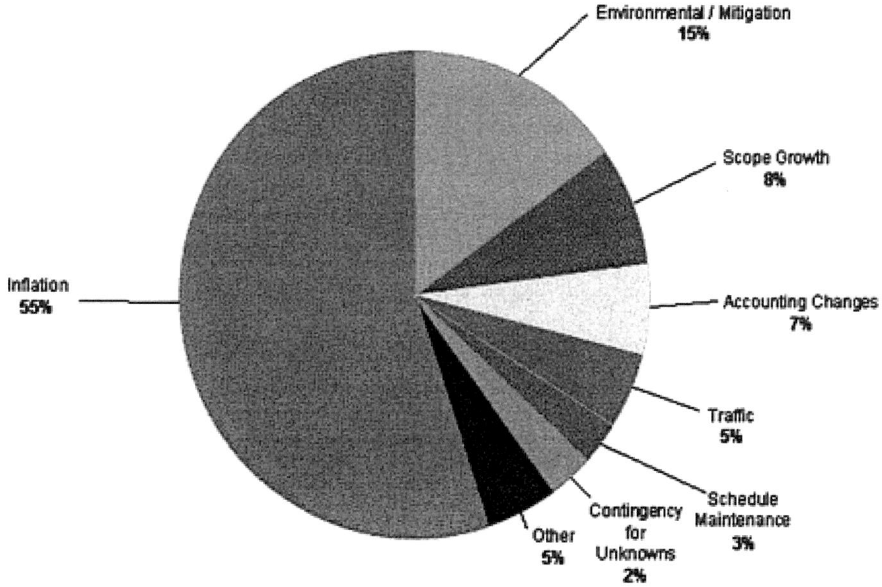

FIGURE 2.1 Cost History and Scope Evolution (Source: W.Edwards, "Project History," Presentation to the Committee on October 21, 2002)

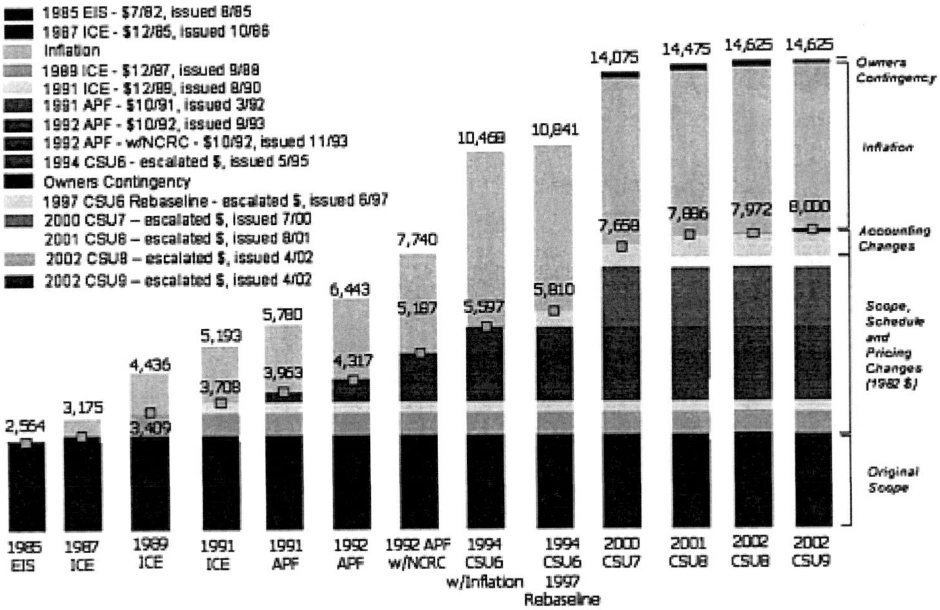

FIGURE 2.2 Cost History and Scope Evolution, in millions of dollars (Source: W.Edwards, "Project History," Presentation to the Committee on October 21, 2002)

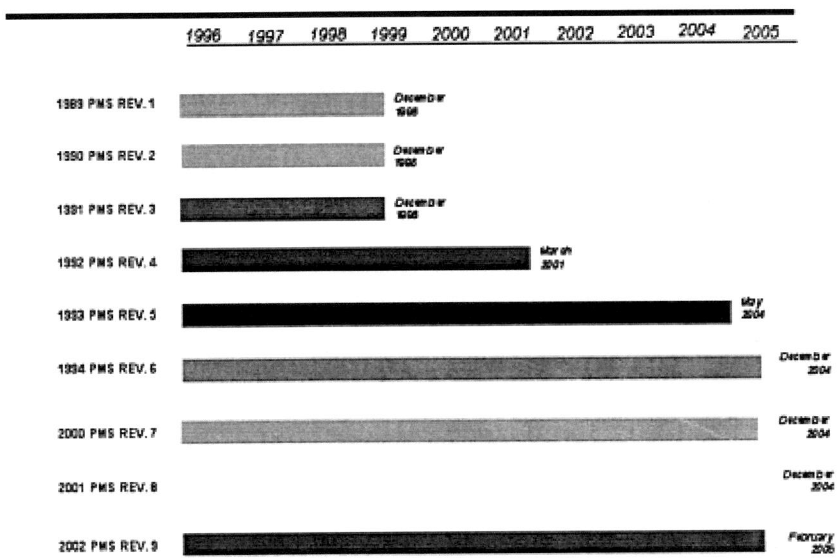

FIGURE 2.3 Project Schedule Escalation (Source: W.Edwards, "Project History," Presenation to the Committee on October 21, 2002)

Downtown Boston is the location of many of America's most important educational, cultural, and financial institutions, and the project extends into several residential neighborhoods as well. Mitigation costs, therefore, could be expected to be very high. Requirements for significant mitigation were identified in the original 1985 EIS, but the costs were not identified.

The project team described mitigation resolution as a consensus-building process involving negotiations with neighborhood, business, development, and environmental groups to ameliorate negative impacts of the construction. This occurs, for instance, by introducing major design changes, routing the construction around residential areas, and constructing bypass roads, all at considerable cost. There is by no means any universal agreement, however, that the money is always spent well or in the service of stated goals. The Honorable Tom Petri (R-Wisc.), chairman of the House Transportation and Infrastructure Committee, has called CA/T expenses that had little or nothing to do with construction a "golden opportunity" for spending federal highway funds on local government agencies, businesses, and educational institutions (Petri, 1997).

Even without the advantage of hindsight, the risks of highly technical engineering design and construction, unknown soil conditions, existing underground utilities, and other factors that increase costs and delay schedules should and could have been anticipated and addressed through additional planning and design, site analysis, and cost and schedule contingencies. As it happened, the impact of unanticipated physical factors was often compounded by delays in making timely project decisions. This was the case, for example, in the delayed decision making on the Charles River crossing, which caused years of delay on this part of the project.

COST AND SCHEDULE CONTROL

Value Engineering

Two types of value-engineering programs—management-initiated and contractor-initiated—have been employed on the CA/T project. The project-management team presented an example of how a management-initiated design change ofhternal tunnel finishes was used to control costs. Reducing the height of the tiled surfaces, using stainless steel panels, and eliminating tiled trim at the junction of the wall and ceiling saved an estimated $70 million. However, the stainless steel panels have apparently interfered with the wireless communications system, creating unanticipated problems not yet resolved at the time of this writing. In any case, the IPO indicated that the $70 million saved on this change was only a part of $480 million saved overall through management-initiated value engineering.

Another example presented by the integrated project organization (IPO) was the Fort Point Channel crossing, which was said to benefit from significant engineering and management effort to control cost and schedule. However, the committee questions whether there was sufficient effort to explore the alternative approach, a bridge crossing, which would have been much less expensive.

The committee was informed that all construction contracts had a provision, called the value engineering change proposal (VECP) program, for contractor-initiated value engineering. The project team claimed savings of $25 million from this program thus far, though the committee believes this is negligible when compared to the $9 billion value of construction.

However, the total savings from both VE programs is approximately $505 million, according to the project staff. This value is approximately 5.8 percent of the total construction commitments to date, which is slightly above the average for VE savings on federal aid highway programs (FHWA, 2002).

As noted earlier, the design of the project is virtually 100 percent complete and the construction is 85 percent complete, thereby limiting opportunities for additional cost and schedule savings from value engineering changes. Nevertheless, the committee believes that the VECP program should be pursued.

Fast-Track Construction

The project used a fast track design and construction delivery method to reduce the overall project time. Typically, fast-track delivery involves dividing the project into sequential work packages and starting construction on the earlier segments before completing design of the latter segments. This process works best when the design of each work package is complete and the impact of requirements of the later segments can be anticipated. The CA/T work packages had complete civil design, but they frequently required modifications to accommodate project-wide systems that were designed in later packages. The result has been a high rate of claims and changes, which is discussed in Chapter 3.

An analysis by the project-management team claims that the project has realized a 3:1 cost-benefit ratio by utilizing fast-track methods (i.e., the accelerated schedule saved $3 for each $1 of additional costs attributed to using fast track). However, the committee does not believe that the full cost impact of changes, claims, and delays from incomplete coordination of work packages has been considered. It should be noted that different people interviewed by the committee defined "fast track" differently, which further complicates the attribution of benefits.

Managing Cost and Schedule

Attempts to mitigate cost increases and schedule delays have been made through application of project-management tools. The project management consultant Bechtel/P arsons Brinckerhoff (B/PB) developed a dynamic Milestone Manager process to provide real-time project performance data; B/PB used it to forecast project delays and develop new work sequences to work around the delays. For example, when I-90 was 19 months behind schedule, the consultant was able to recover 8 months of delay by accelerating other key parts of the project.

CURRENT STATUS

With the project so close to being complete, the risks of future scope increase, mitigation requirements, and technical problems should be reduced and there should be a high level of certainty on the remaining cost and time. The budget includes substantial cost, schedule, and administrative contingencies, and the staff has almost 20 years of project experience all of which contribute to minimal decision risk. Nevertheless, the project continues to experience delays and costs increases, most recently exemplified by the delayed opening of the Interstate 90 connection to Logan Airport from September 7, 2002, to November 8, 2002, then to December 20, 2002; the connection finally opened over the weekend of January 18, 2003.

The potential cost of that delay was covered by the budget contingency, but the committee believes that delays in such critical milestones at this stage of the project are unacceptable and can severely undermine the credibility of the MTA. Therefore there is a need to identify and assess the current cost and scheduling risks and to develop strategies to manage them through completion of the project.

The IPO reported to the committee that the continued delays in opening the I-90 extension were primarily caused by one contract that involved the installation of traffic monitors and controls in the tunnel. The IPO also reported that it was accelerating work on all contracts to meet the scheduled completion date, but that the date was still continuing to slip. The committee concludes that the persistent scheduling problems result from failures in coordination together with insufficient emphasis on accelerating work that would be most critical to maintaining the schedule.

The Washington State Department of Transportation (WSDOT) has developed a Cost Estimating Validation Process (CEVP) that resembles value engineering. In this procedure, an independent team of experts is brought together in a workshop setting to review the project's estimated cost and schedule, as well as its risk-management strategies. The team not only identifies actions that will improve performance but also helps communicate to the public the assumptions and potential cost impacts of identified risks (WSDOT, 2002). The CA/T could have benefited from a CEVP-type process when it began in 1982, and although options are now limited such a process could still help the IPO to control cost and schedule for completing the project (see also the discussion of peer reviews in chapter 5).

FINDINGS AND RECOMMENDATIONS

Schedule Management

Finding 1: Meeting the CA/T project's scheduled targets continues to be a problem. Despite an emphasis on reaching the milestones on time, slippage continues, thereby reducing public confidence in CA/T management.

This problem was particularly evident during the continual postponement of the opening dates of I-90 and I-93N, which occurred while the committee was doing its work. The slippages seem to stem from managers focusing more on the short-term details of the project activities and less on evaluating the project risk as a whole in advance of potential occurrence.

Recommendation 1: CA/T project managers should strategically evaluate future schedules by determining what critical tasks need to be done without fail and how long these activities will likely take. Published completion dates should be developed around realistic workflows and schedule risks, with modest allowances for unknown issues.

These evaluations should allow the project director to make reasonable projections for completion of the project. The work could then be pushed to deliver even earlier, thereby possibly beating the published date.

Cost Management

Finding 2: The project has had large cost increases resulting from changes in scope, design, and project limits, as well as from deficiencies in coordinating contracts.

Recommendation 2: All contract modifications should be comprehensively reviewed— prior to execution—for impacts on scope, design details, interfaces, and contract duration.

Reliable cost and schedule estimates should be based on consideration of alternatives, full development of scope, knowledge of existing conditions, sequence of activities, and the coordination work between packages.

Cost/Schedule Tradeoff

Finding 3: The committee could not identify sufficient analysis to justify all construction- acceleration efforts that have occurred.

There is a concern that the premium costs for construction acceleration (extended periods of overtime and longer workweeks, for example) could lead to problems with work quality and a bigger-than-expected list of items to be reworked or completed prior to acceptance.

Recommendation 3: The MTA should rigorously analyze all costs associated with project acceleration, especially for completing the key critical path tasks (those in which a delay could cause critical project milestones to be missed).

Value Engineering

Finding 4: The savings from contractor-proposed "value engineering change proposals" (VECPs) to date is $25 million, which is a negligible amount compared with the constructed value of the project.

Recommendation 4: The MTA should aggressively promote the VECP program for the remaining work. This effort should increase contractors' level of awareness, encourage innovation, and yield cost reductions.

REFERENCES

Flyvbjerg, B., M.S.Holm, S.Buhl. 2002. "Underestimating Costs in Public Works Projects: Error or Lie?"*Journal of the American Planning Association.*68(3): 279–295.

Campbell, Robert. 2000. "Forget overruns; Big Dig's a marvel."*Boston Globe.*25 August, pp. C1 and C8.

Wachs, Martin. 1989. "When Planners Lie with Numbers."*Journal of the American Planning Association.*55 (4): 476–79.

Merrow, Edward W.1988. *Understanding the Outcomes of Megaprojects.*The Rand Corporation.

Petri, Tom. 1997. "Big Dig Wins Second Porker." Available online at http://www.house.gov/petri/porker/sepl9prk.htm. [January 20, 2003].

Federal Highway Administration (FHWA). 2002. "Summary of Past VE Savings—Federal Aid Highway Program." Available online at http://www.fhwa.dot.gov/ve/ [January 22, 2003].

Washington State Department of Transportation (WSDOT). 2002. "Cost Estimating Validation Procedure." Available online at http://www.wsdot.wa.gov/projects/cevp/default.htm [December 30, 2002].

3

Contract Administration

INTRODUCTION

The 20-year span of the CA/T project covers a period of change in the administration of public-works projects in the United States. During that time, the previous and long-standing approach of adversarial contracting evolved into what has been described as "modern contracting practices," which are now being used by an increasing number of public owners (Hancher, 1999; Smith, 1993). These practices include explicit consideration of risk allocation; partnering among owner, designers, and construction professionals; alternative dispute resolution; alternative contracting methods such as Design/Build and contracts including charges for closing roadways (commonly called "A+B" contracting); emphasis on quality, safety, and scheduling; and value-engineering change proposals. Altogether, the evolved practices include sophisticated processes for management and administration of design and construction as well as for cost and schedule reporting systems. Consequently, the public owner needs to provide a highly educated and experienced staff to manage and administer the projects and to protect the public interest.

CONTRACTING AND MANAGEMENT PRACTICES

The CA/T project has employed a number of modern contracting practices, including alternative dispute resolution, that successfully resolved more than two-thirds of the project's 9,163 contract modifications. For example, the committee was informed that construction contracts with a value greater than $20 million utilize dispute-review boards, and that mediation was being used on several contracts. A review of Massachusetts's *Boilerplate Construction Contract Document* showed provisions for differing site conditions; value-engineering change proposals (VECPs); voluntary partnering and dispute-resolution boards; and provision of geotechnical information. State regulations preclude the CA/T project from using alternative contracting practices such as Design/Build and A+B.

The CA/T project also has incorporated a wide variety of procedures to support the administration of the project contracts. The committee reviewed a number of these, including those documented in the *Project Engineers Manual; Design Policy Memorandum List and Cross References; Partnering Manual; CA/T Project Procedures;* and CA/T *Construction Contract Administration Manual* (MTA; MTA, 2002a; MTA, 1998; MTA, 2002b; MTA, 2002c). Additionally, over the course of the study, information about additional techniques being used for contract administration—including meetings, reports, audits, and oversight—was presented to the committee.

CLAIMS AND CHANGES

Claims Management

Settling changes and claims on the project have constituted a significant effort. The term "claim" is used by CA/T management to include all change requests submitted by contractors, including those that are resolved and not contested. The term also covers matters that have not yet been agreed to and could be disputed. The committee was told that as of December 2002, there were approximately 3,200 unresolved construction claims and 350 unresolved engineering claims, with a combined value of $230 million and an average age of over 600 days. The number or cumulative value of the claims that will require actions beyond negotiations and the degree to which mediation or litigation will be necessary to resolve them are not known.

It is clear that the exposure of the project to those claims that have been pending for long periods of time is especially significant if interest is included. This is compounded by the increasing difficulty of finding documentation and recollecting details as the issues age, as well as by loss of institutional memory through staffing reductions as the project nears completion. This is already occurring, as a key project staff member currently involved in changes and claims management is leaving.

In May 2000, the MTA chairman assigned responsibility for claims and changes of over $250,000 to the MTA legal staff. In the committee's collective experience, it is the nature and culture of legal staff to favor litigation for resolution of issues—an approach that is not typical of the modern construction industry and not inclusive of the field people who have technical construction backgrounds and knowledge of the issues (TRB, 1998). Therefore this current MTA approach runs the risk of focusing primarily on litigation and may obstruct the use of such standard industry practices as alternative dispute resolution.

The project also uses dispute resolution boards (DRBs) to deal with claims on larger issues and contracts. These DRBs issue nonbinding rulings, consistent with the recommendations of the Dispute Review Board Foundation, so neither party is forced to accept the judgment. Some contractors who participated in the committee's roundtable discussion suggested that the nonbinding nature of the process serves only to benefit the owner.

Modifications

Historically, changes and claims for the CA/T have been settled for approximately 50 percent of the contractor's original claimed cost. The IPO estimates that payments for changes and claims have totaled $1.6 billion—9,163 modifications with an average value of $175,000— worth approximately 20 percent of the construction commitments as of October 2002. The committee found it difficult to benchmark these numbers, however, considering the uniqueness of the project from the standpoints of total cost and complexity. There is additional exposure on the $2 billion in construction yet to be completed, and contingency amounts are budgeted to cover these costs.

The committee was provided with a breakdown of the categories of contract modifications. Figure 3.1 illustrates that "Schedule Adjustment" and "Scope Transfer & Change" accounted for some 29 percent, or $473 million, of the $1.6 billion in modifications to date. "Design Development" accounted for 16 percent, or $260 million. These costs reflect the difficulties encountered in managing the issues associated with incomplete design, schedule issues, and contract interfaces. The committee notes, for example, that scope modification bears a cost—the price of a negotiated change for additional work is greater than the credit that can be negotiated for deleted work. The degree to which contingency has funded scope additions was not specifically identified, although there are categories such as "Other" (28 percent, or $460 million), where it may have been included.

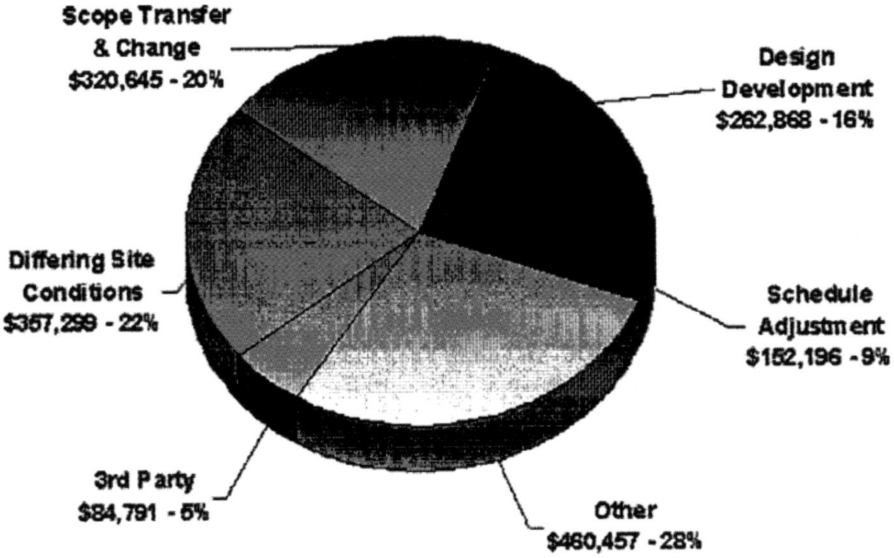

FIGURE 3.1 Modification Cause by Dollar (Thousands) (Source: J.Gorman, "Construction Claims and Changes," Presentation to the Committee on November 11, 2002)

The MTA presented to the committee an extensive process now in place for managing claims and changes. The foundation for this is a set of goals and objectives from which procedures and methodologies have been developed and implemented.

Cost Recovery

The design of the CA/T is largely complete, and project managers are now working to recover costs under the project's error and omission review program. This also involves contractor backcharges. If a design-engineering firm were liable, it would be required to reimburse the project under its professional liability coverage. But because the project has an owner-controlled insurance program (OCIP), it is possible that this effort could result merely in moving funds from one account to another within the project; making it a symbolic rather than a true cost saving.

CONTINGENCY

Contingency funds are now budgeted for claims and changes in three categories: construction contracts, management contingency, and MTA CEO contingency. The last two categories are funded at $294 million and $40.9 million, respectively. The CEO contingency is controlled exclusively by the MTA chair and is used specifically to offset the added project cost of MHD staff and to expand the claims and changes administration.

Within the construction contracts category there are allocations for quantity overruns and underruns, issues and claims, and future allowance. The committee was not presented with specific amounts for these line items. However, their total amount (including two others—police details and unawarded) totals $955.5 million. It would thus appear that the total contingency available is in excess of $900 million for claims and other issues. The most recent audit of the project, in September 2002, indicated that this amount of contingency was sufficient (Deloitte and Touche, 2002).

Industry guidelines exist for the development of contingency budgets and their use. For example, the Project Management Institute has extensive information on contingency development and management (PMI, 2001). In general, the process begins with a risk analysis to develop issues that will enable the quantification of a project-contingency allowance. As the project moves through planning, predesign, design, and construction, the contingency percentages are refined because more is known about the project. Industry practices recognize that contingency funds might be required for such things as "items omitted" because of factors such as lack of definition, abnormal conditions, underestimated quantities, changes in regulations and codes, and environmental requirements. A risk analysis is typically used to quantify each of these types of items.

Contingency funds are not intended to be used for scope additions or other project enhancements, which should be justified and funded through a separate budget process and receive approvals from top management or a policy board. The challenge for the CA/T project managers will be to use the contingency funds for their intended purpose and at a level that is appropriate for the remaining work. The objective should not be to spend all contingency funds but rather to use them prudently, perhaps reducing the total project cost. The committee is not aware of any inappropriate uses of the contingency funds, but it wishes to state clearly that the availability of contingency funds should not be viewed as an opportunity to fund enhancements or add scope to the project.

FINDINGS AND RECOMMENDATIONS

Claims Management

Finding 5a: The project's exposure for the unsettled claims and changes is significant. Both the number of claims and their average age have been growing steadily for the past 4 years.

As of December 2002 there are more than 3200 open construction issues and 350 engineering issues, with an average age of more than 600 days. This large number of unsettled changes and claims also significantly affects the contractors, as many have substantially completed work for which they have not yet been paid. No documented plan was presented for dealing with this backlog.

Finding 5b: Responsibility for resolving claims and changes for amounts over $250,000 is currently assigned to the legal staff of the MTA, and the imminent departure of a key staff member may adversely affect staff productivity.

Recommendation 5a: Resolution of unsettled changes and claims should be deemed a high priority issue, meriting immediate attention by the MTA. It constitutes a project within the project, to which project management principles should be applied.

A special group should be created within the project organization to manage these issues and others related to contract closeout. The group should be located within the construction arm, with support from legal staff, headed by an experienced construction manager, and staffed with experienced closeout managers. This group should have full authority for carrying out its responsibilities, along with the ability to obtain the necessary support. Closeout managers should be recruited from within the existing project organization in order to retain institutional memory, and outside personnel with special claims experience should also be considered in order to bring a broader perspective to the closeout process.

Recommendation 5b: A closeout schedule should be developed, along with identification of resources necessary to support it. A target date of July 2004 should be established for resolution of the unsettled claims and changes.

Efforts should be made to resolve future issues on a timely basis so as to avoid an even greater accumulation of unresolved issues and costs. Timeliness and fairness, along with protecting the public interest, should be the overarching objectives.

Recommendation 5c: Consideration should be given to paying contractors for the estimated direct costs of directed changes and recognized claims in advance of settling payment for other indirect cost that they may be entitled to so that the burden upon contractors and subcontractors carrying these costs can be relieved.

MTA and the contractors should agree to a final settlement plan for dealing with other indirect costs as a condition for this earlier payment. Alternative dispute-resolution methods offer the opportunity to provide timely closeout of claims and to reduce their impact on the project and contractors alike.

Contingency Management

Finding 6: While individual contracts have been analyzed for exposure to changes, comprehensive risk- and contingency-management tools and processes do not appear to be in place.

The project now has budgeted for contingencies in three separate categories: construction contracts, management contingency, and MTA CEO contingency. As of December 2002, it appears that the total contingency available is more than $900 million. If these funds remain unused for their intended purpose as the project nears completion, they should not be viewed as a funding opportunity for enhancements or added scope. Rather, they should provide an opportunity for reduction in the total project cost.

Recommendation 6: The MTA should initiate contingency-management procedures that include: a continual comprehensive risk analysis to quantify and refine contract contingencies, individual contract contingency tracking, and a contingency drawdown plan that includes contingency-use forecasts.

Cost Recovery

Finding 7: The current cost-recovery program emphasizes legal issues prior to resolution of engineering issues. This process is reported to have replaced a partnering-style relationship with an adversarial process.

The program is designed to recover costs from design engineers that could be attributed to errors and omissions in the detailed design of the project. MHD developed the program originally in 1993. The latest program revision (Revision 5) was developed in March 2002 by MTA and places authority to pursue claims with the MTA Chairman. It differs from earlier programs in that the initial evaluation of potential cost recovery has been shifted from the technical working level of the project to staff attorneys and outside legal claims-management consultants. The revised program also combines the cost-recovery processes for design and construction, which presents the risk of confusing engineers' and contractors' respective performance standards, levels of risk, and financial responsibilities.

The first $500,000 of design-cost recovery would come from the design engineer or its insurance carrier, and the amount beyond that would come from the project OCIP. At the present time the OCIP's level is $20 million, which is well below the intended $50-million level. This gap results from the financial problems of one of the companies underwriting the OCIP, program so the MTA is currently looking for another insurance underwriter. Meanwhile, the MTA does not expect this temporary setback to threaten the financial health of the project.

Recommendation 7: The MTA should take steps to reinstate the engineering-driven process in order to resolve cost-recovery issues.

An effort should be made to resolve technical issues before initiating an adversarial process. Additionally, alternate dispute-resolution techniques should be considered to help resolve these issues. The cost-recovery program for design professionals should be well communicated to the affected firms so that there is a clear understanding of the issues and their impacts on the project.

REFERENCES

PMI (Project Management Institute)2000. "Guide to the Project Management Body of Knowledge."Project Management Institute, Inc.Newtown Square, PA.

Hancher, D.E.1999. "Contracting Methods for Highway Construction."*TR News*. Transportation Research Board. 10–15.

Smith, R.J.1993. "Using Modern Contracting Practices for No-Dig Projects."*Proceedings of North American No-Dig '93*.NASTT. Paper A1.

Massachusetts Turnpike Authority (MTA). *Project Engineer's Manual*.Revision 4.

Massachusetts Turnpike Authority (MTA). 2002a. *Design Policy Memorandum List and Cross References*.October.

Massachusetts Turnpike Authority (MTA). 2002b. *Central Artery/Tunnel Project Procedures*. August.

Massachusetts Turnpike Authority (MTA). 2002c. *Construction Contract Administration Manual*,updated August.

Massachusetts Turnpike Authority (MTA). 1998. *Partnering Manual*.January.

Transportation Research Board (TRB). 1998. "Managing Transit Construction Contract Claims."TCRP Synthesis 28. Washington, DC.National Academy Press.

4

Reporting and Controls

INTRODUCTION

The processes for reporting and controls have changed significantly over the course of the CA/T project. During the 1990s a comprehensive project database was developed, using state-of-the-art relational database software, to enable the detailed tracking of individual contract tasks and the assembly of cost reports. Since no off-the-shelf programs were capable of handling such a mammoth project as CA/T, management consultant Bechtel/Parsons Brinckerhoff (B/PB) developed a formidable construction-information system (CIS) using a commercial software system. The CIS, which was not fully developed and available for use until about 5 years after construction began, tracks almost every aspect of the project. That includes all design and construction contracts and their funding sources, requests for information, change orders, submittals, design specifications, accounting and budgeting, forecasting, staffing, scheduling, and reporting. However, there continued to be major problems—in the flow of information to oversight groups, in the provision for adequate contingency, and in the inconsistency between project scope and cost estimation. In 2000 the MTA announced a substantial project overrun, to the apparent surprise of the FHWA oversight group. Following that event, the reporting and control processes were modified to track budget and schedule progress to a fixed project scope.

Using the CIS, the project team generates a large number of progress reports, providing what the committee believes might be an excessive amount of information (6,400 data tables, 167,000 columns) for managing the project. Much of that information lacks the strategic view needed to develop a reliable construction schedule, though this information may serve a purpose during the operation and maintenance of the highway system.

REPORTING

Project reports for various time intervals are regularly scheduled. A progress report on each active contract is circulated daily via email to resident engineers, project managers, and oversight agencies in order to summarize work progression, quality deficiencies, and identified needs for modifications. These daily reports are based on the detailed work logs maintained by the resident engineers.

A monthly set of reports is produced to identify exceptions (or variances) to the existing budget and schedule. In addition, a set of month-end documents for each contract is developed and reviewed by senior management and oversight agencies to inform them of issues associated with contract modifications and claims. For example, a *Budget, Cost, Commitment and Forecast* report is prepared for financial review by oversight agencies. In addition, a *Project Management Monthly* report is prepared for senior management, oversight agencies, and the public. It summarizes the status of budgets, cost exposures, schedules, safety records, and employment-diversity records for the entire project, but it does not include detail on the status of individual contracts.

A major annual report for the project is the *Cost and Schedule Update* (CSU) prepared for oversight agencies and bondholders. Since 2000, the accounting firm Deloitte & Touche has conducted independent reviews of the CSUs on behalf of the Massachusetts Executive Office for Administration and Finance. The Federal Highway Administration also reviews the CSUs and prepares independent estimates of completion costs and schedules.

CONTROLS

Cost control relies on a resource-loaded work-breakdown schedule. Each contract is divided into specific tasks with required resources and schedule milestones, and detailed information on specific tasks can be pinpointed or aggregated. As contract modifications are implemented, additional tasks may be added to the schedule and the work-breakdown structure.

The implementation of the IPO has complicated the control of expenses for the B/PB management-consultant team. MTA personnel, including the director of finance, review and approve invoices from the consultant. Annually, the staffing of the B/PB management consultant team is reviewed, with the expectation that time commitments will be reduced as the CA/T project concludes more contracts and moves toward completion.

Schedule control is maintained by monitoring work-in-progress and planning future tasks. Contractors develop detailed schedules for their own work that are aggregated into the overall project schedule, and milestones are set for individual contracts and the project as a whole. Scheduling derives from a commercial software package, Primavera™, in widespread use in the construction industry.

This software is being pushed to its limits—the current schedule for completion of the CA/T can be regarded as tight, with numerous overlaps in activities for finishing work and aggressive estimates of completion time for operational requirements—and in any case cannot compensate for problems on the work site. Indeed, the milestone date for opening I-90 was extended 81 days and the milestone date for the initial I-93 Northbound opening was extended 53 days between the forecast in CSU 8 (August 2001) and in CSU 9 (July 2002) (Deloitte & Touche, 2002).

Control of the project scope is a critical issue on a large, complicated project such as the CA/T. Since the original \$2.6 billion estimate (in constant 1982 dollars), numerous additions have been made, especially for mitigation of environmental or economic impacts and to adjust for unanticipated site conditions. Currently, the senior project management reviews significant scope changes as part of the contract-modification process. As the project enters into its final stages, however, the committee expects that changes in scope should be less and less of an issue.

OBSERVATIONS

Project reports since 2000 have been consistent with state-of-the-practice on large construction projects, though the committee believes that the CA/T would have benefited greatly if the above-mentioned reporting procedures—especially the annual cost and schedule updates and the improved information flow to oversight agencies and the public—had been implemented much earlier. The committee notes, however, that the public-information website appears to be a useful and effective outreach tool.

The CA/T engineering design and construction functions are separate, without a project manager responsible for the completion of distinct components. This lack of an overarching project-management organization hampers control and the transition between functions; it may well have contributed to the large number of required contract modifications.

Scheduling continues to be a problem on the CA/T project, as evidenced by the fact that milestones continue to slip. Scheduling might be improved by adding some flexibility, with tracking for "optimistic," "most likely," and "pessimistic" estimates. Also, wider use of incentives tied to task completion could improve schedule control.

FINDINGS AND RECOMMENDATIONS

Information Overload for Senior Management

Finding 8: Too much uncoordinated information is provided to the project's senior management, and project reports do not lend themselves to ready analysis—particularly with regard to such concerns as the uncertainty of forecasts and the trade-off possibilities between time reductions and cost increases.

The CA/T project staff routinely generates extensive cost and schedule reports, conducts regularly scheduled meetings, and is able to provide extensive information on a myriad of project details. Special reports can be developed on request, as evidenced by several special studies developed within a day for this NRC panel. Summary information and strategic issues, however, need to be emphasized. For example, the *Project Management Monthly* report provides good information about the process of construction, but it is less effective for contingency management or identification of critical issues.

Recommendation 8: As the project foci shift—to managing claims, cost-recovery processes, and the transition to operation—project reports should also change to emphasize summary information and strategic issues.

Reports should focus management attention on critical management issues such as contingency management and critical-path schedules. Summary information and strategic issues need to be emphasized in the project reports.

Schedule Analysis

Finding 9: The MTA has developed an aggressive and optimistic schedule for project completion. However, the project milestones are prone to slippage as events occur.

The project's monthly reports include milestone dates and contingency dates, but they do not necessarily offer realistic completion dates. This slippage creates problems of public credibility for the CA/T management and the MTA.

Recommendation 9: Project reports should reflect the probability of meeting milestones or the probabilistic distribution of completion times, based on the results of past efforts.

Project reports include schedule milestones and the budgeted contingency milestones, but they provide little help to reviewers regarding the likelihood that these milestones will be reached on time. Additional information, such as optimistic, pessimistic, and most likely milestone-time achievements, would be helpful. These estimates could be based on the scheduling experience garnered over the course of the project.

REFERENCES

Deloitte & Touche. 2002. "Central Artery/Tunnel Project: Project Assessment."Prepared for Massachusetts Executive Office for Administration and Finance. September 30, 2002.

5

Oversight of Management Consultant

INTRODUCTION

The CA/T project now operates with an integrated project organization (IPO) comprised of staff both from MTA and the management consultant Bechtel/Parsons Brinckerhoff (B/PB). Responding to the challenge of designing and managing one of the largest public-works projects in United States history, this structure evolved to combine the expertise of the management consultant with the longer-term dedication and specialized experience of the owner. This organizational structure exists because of the knowledge that the owner has acquired of the strengths and weaknesses of the staffs and the trust that has developed from years of working together on this project. By most accounts, the IPO appears to be operating reasonably well in this phase of the CA/T project.

During the transition to the IPO structure in 1997–1998, the best-qualified person available for a particular managerial position was selected regardless of organizational affiliation. (The position of project director, who reports to the chairman of the MTA, was reserved for an MTA employee.) In effect, B/PB is no longer in the role of a project management consultant but supplies highly qualified people to augment the staff of the MTA.

This arrangement does not resemble the usual owner/consultant relationship, in which public-sector employees and the consultant's employees operate independently. Some of the reasons for this traditional structure have been to establish clear lines of responsibility and accountability, to create a system of checks and balances, to eliminate perceived conflicts of interest, and to have a clear understanding of owner/consultant risk allocations.

At the beginning of the CA/T project, the relationship between the state and the consultant reflected this usual owner/consultant relationship. However, as the project moved from design to construction, the structure was modified to address performance problems, enhance communications, reduce the number of management layers, and moderate conflicts between owner and consultant.

Another transition is currently needed, however. With engineering essentially done and construction completion only three years away, the need for the consultant's staff is diminishing rapidly. Yet the committee observed that a detailed plan for downsizing and elimination of the management consultant staff was not in place.

OVERSIGHT

The MTA project director has the responsibility to monitor and oversee the management consultant. Contractor oversight is accomplished in large part through meetings and key metric reports such as:

- Weekly interface meetings,

- Internal review *of Project Management Monthly* reports,
- Weekly critical issues meetings/reports,
- Monitoring month-end close results,
- Review of monthly *Budget, Cost, Commitment and Forecast* reports, and
- Daily progress reports and follow-up.

In addition, the project director receives feedback from the FHWA staff who are involved in the daily monitoring and oversight of the project.

PEER REVIEW

Owners and project managers of complex engineering projects often use independent peer review as a means of quality assurance. Project managers recruit qualified professionals with related experience to review the project team's assumptions, procedures, and decisions. They challenge the project team members to defend their work and in the process rethink decisions to assure that the optimal choice has been made.

Peer review, which essentially provides an outside perspective to identify issues that may have been missed, has been used to review technical decisions as well as project-management decisions. For example, the U.S. Department of Energy has used this process to address problems on projects ranging in size from $5 million to several billion dollars (NRC, 1998). Independent advisory boards have also become customary for projects, such as large dams, that could seriously impact public safety.

The committee was told that the CA/T project team conducted peer reviews. However, they were undertaken by employees of the joint venture organizations from offices outside the Boston area and thus were not truly independent.

At this point most, if not all, of the technical-design decisions for the CA/T project have been made, which essentially renders moot the value of technical reviews. However, with almost $2 billion in construction still to be completed, a focused independent peer review of project management decisions could yield meaningful savings.

FINDINGS AND RECOMMENDATIONS

Integrated Project-Management Oversight

Finding 10: The integrated project organization (IPO) structure currently utilized by the MTA to direct the design and construction of the CA/T project appears to be functioning reasonably well. Although most of the contracts are now under way or completed, the committee observed that a detailed plan for downsizing and ultimately eliminating the management consultant staff was not in place.

Recommendation 10: MTA should implement an aggressive plan to downsize the B/PB staff members who are not essential to completing contracts and resolving claims. At the same time, the MTA should ensure that key staff members remain in place to finalize and close out all contracts and claims.

The focus of the IPO over the remaining life of the project should be (1) completion of all contracted construction according to a realistic scope, realistic schedule, and the existing budget; and (2) cost-effective resolution of all claims and open issues.

Finding 11: The existing management tools and metrics currently in use are sufficient for overseeing the management consultant.

Recommendation 11: If the MTA wishes to improve its oversight of the management consultant, the committee encourages it to make better use of the tools presently available rather than develop new tools.

Independent Peer Review

Finding 12: The CA/T project has had numerous reviews over time for different purposes. However, there has not been a continuing independent peer-review of technical issues, management, and strategic directions.

Many organizations have formed technical advisory boards to review their work and periodically provide advice. The MTA Board fills this role to some extent, but it does not have direct involvement in the CA/T project and lacks the range of experience desired in a technical advisory board. From its inception, the CA/T project extended over more than two decades and would have benefited from an independent technical advisory board. In particular, such a board could have aided oversight in identification of operational and technical issues. But it's not too late. A technical advisory board could still be extremely helpful during the remaining life of the project. And with a corresponding shift in membership, the board could also help in the transition to operations.

Recommendation 12: MTA should establish an external, independent peer-review program to address technical and management issues until transition to operations and maintenance is completed. The frequency of reviews and peer-review participants should vary in accordance with the issues to be addressed.

REFERENCES

NRC (National Research Council). 1998. *Assessing the Need for Independent Project Management Reviews in the Department of Energy.* Washington, D.C.: National Academy Press.

6

Transition from Construction to Operations

USER TRANSITION

After a 20-year process of planning, designing, and constructing the CA/T project, the final major transition will be to operations and maintenance. The public, having long endured the project's disruptions and inconveniences, now expects results—specifically that the new roadways will solve the old congestion problems, making hours of delay while stuck in traffic a thing of the past. But without careful planning, that "moment" could be traumatic. While the owner can grow with operations experience gained over time, users must adapt immediately as they traverse the new system. More than a third of the CA/T project is underground, where motorists do not have environmental factors to give them locational clues; they are totally dependent on the signs, stripes, and lighting along the roadway. Because these clues must be familiar and effective right from the beginning—the motorist's first impression of the new system will likely be a lasting one—the job of communicating with the public cannot be neglected or left to the last minute.

Public Education

The public-education element of the start-up process is critical and should be a major focus of the transition team. If the owner assigns responsibility for it to a nontechnical individual or firm, a technical person will be absolutely necessary to review what is being presented and assure that it is user-friendly and consistent with the reality of the project.

The message should start early and be repeated often, virtually to the point where the public gets tired of hearing it. There is certainly a saturation point for information, but until then the population is slowly absorbing the news. A corollary is that innovative ways of getting the public's attention are necessary, as most people filter information and consciously take in only what piques their interest. All forms of local media must therefore be engaged. Service clubs, auto clubs, neighborhood groups, and the Massachusetts Registry of Motor Vehicles can also help spread the word. And in addition to the average citizen, other important players—the enforcement community, the courts and judges, and traffic reporters—also need to be educated. Needless to say, the message content should be relevant and accurate right from the start.

Unfortunately, information supplied to the committee indicates that the MTA is substantially behind schedule in educating the public and preparing motorists for the new system, and this is a cause for concern. The committee believes that proper education of the users cannot happen with a sudden media "blitz" but will need a carefully planned and targeted stream of information executed well ahead of start-up.

Safety

The committee also believes that motorist safety will require special attention when the CA/T is opened. Because a large portion of the project is underground, with minimal shoulders for disabled vehicles, a stranded motorist may be placed in a hazardous position. The current operational procedures rely on cameras monitored in the Operations Control Center to spot troubled vehicles, and they allow up to 8 minutes to service the scene. During this time, the motorist in the tunnel may be confused, unaware of what is going on, and vulnerable. Guidance for such situations should be included in the public-education process in order to prevent these kinds of user reactions.

MANAGEMENT TRANSITION

The documentation that the committee received from the MTA contains extensive procedural guidance for the transition process. This guidance pertains to operations and maintenance responsibilities, transfer of ownership, safety certification, and the documentation required.

The process's organization chart shows the MTA chief of operations serving as director of transition. It also indicates that 79 percent (72/91) of the transition staff are employees of B/PB. Meanwhile, existing parallel to the transition organization is the operations organization, with approximately 450 MTA employees. But the process by which the MTA will phase out the transition staff (plus the 600 staff working within the IPO) and phase in the operations staff is not clear.

Two important elements of the seemingly missing strategic link should be recognized:

- Maintenance of the CA/T project's institutional memory during transition will require the cooperation of all the organizations involved in design, construction, and operation.
- Staff changes caused by transition and the time required to familiarize new staff members with management processes and administrative procedures can extend the transition schedule.

A Strategic Approach to Transition

The following three diagrams (Figures 6.1, 6.2, and 6.3) represent the perceived relationships between management functions at different stages of the project transition. They illustrate the relative changes in emphasis among the design and construction IPO, claims processing, and operations and maintenance. The diagrams also indicate the need for "strategic thinking" to coordinate the transition process within MTA and prepare the public to use the new highway system safely and efficiently.

"Strategic thinking" is a process for synthesizing and disseminating project information on scheduled openings, milestones, staffing levels, transition status, cost and schedule, risk, contingency budgets, project performance data, and any other data that affect the project goals. Effective communications among the dozens of federal, state, and local government agencies, as well as organizations representing private interests, need to be facilitated so that each is prepared to fulfill its role in completing and operating the CA/T system. The strategic-thinking process integrates all this information with a broad view of the project and determines the criticality and importance of individual elements. Because information flow is dynamic, the process needs to be highly responsive to any new information. Most important, this function should be able to anticipate and prepare for the next scheduled milestone and, as much as possible, for unforeseen problems.

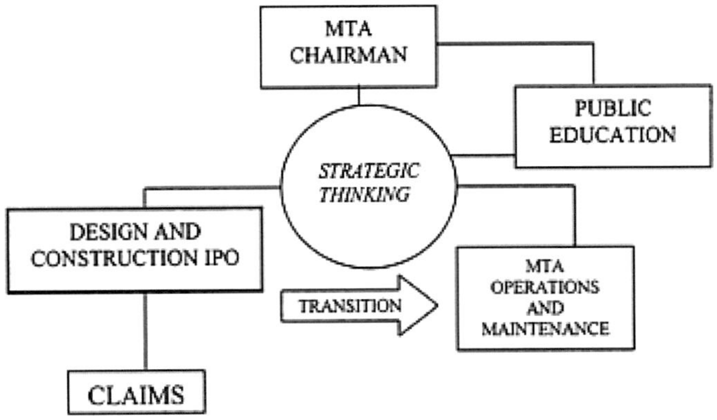

FIGURE 6.1 Transition Phase 1—Completing Construction

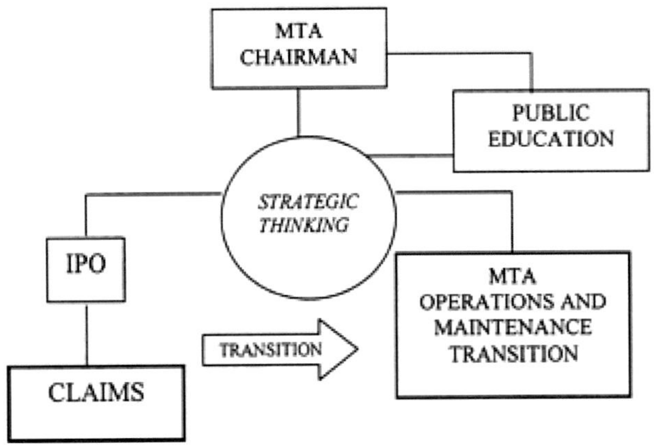

FIGURE 6.2 Transition Phase 2—Transition to O&M and Resolving Claims

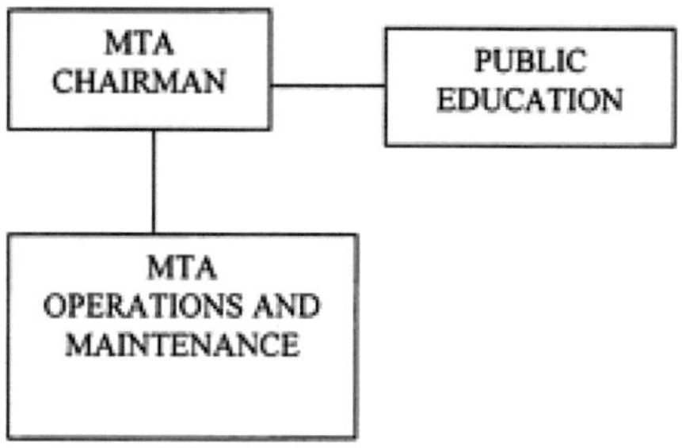

FIGURE 6.3 Transition Phase 3—Operation of system

SECURITY

During the project field review, in October 2002, the committee perceived a lack of security for the CA/T project sites. In these times of heightened need for critical-infrastructure protection, CA/T site security is essential. This should not be a problem, however, given that the available police force, provided to the MTA by the State Police, will double in size during the transition period as part of the operations provisions. By accelerating that increase, this resource could be used for construction-site security until the full demands of operating the CA/T are known.

FINDINGS AND RECOMMENDATIONS

Public Education

Finding 13: Information transfer to the motoring public has been minimal thus far and not well timed for the scheduled opening of the first completed sections of the project.

The committee saw little evidence of the type of media program necessary for preparing the public to use a new, complex, and possibly challenging highway system. The media-relations functions, which are based in the chairman's office, do not have the technical guidance for helping to prepare the needed public information.

Recommendation 13: MTA should have a media team, with technical assistance provided by the project staff, to develop and immediately implement an innovative plan for educating the public to use the CA/T.

The training should include traffic patterns and procedures for normal operations as well as actions to be taken during accidents, breakdowns, and emergencies. The public-education plan should be an integral part of the strategic thinking for the transition of the project to MTA operations and maintenance.

Strategic Thinking

Finding 14: The committee has not seen an adequate plan for guiding the transition from a construction organization dominated by project-management consultants to an operations organization that is largely composed of full-time MTA staff.

The transition from design and construction of CA/T to its operations and maintenance will be a major challenge. The committee believes that the MTA currently lacks strategic thinking for managing this transition process.

Recommendation 14: The MTA should establish processes to assist its chairman in identifying and prioritizing strategic issues and in carrying out the transition from design and construction to operations and maintenance.

The MTA needs processes for responding to changing conditions. These processes should: 1) guide the phasing-out of the IPO; 2) close construction contracts; and 3) maintain controls while the MTA assumes responsibility for the CA/T.

Security

Finding 15: During its October site visit, the committee observed what it perceived as a lack of security.

The committee had signed in and carried credentials for its visit, but it appeared that open access was available at all project-access points. Furthermore, there was no evidence of a security-force presence anywhere in the project area.

Recommendation 15: A comprehensive security program should be developed, immediately implemented, and maintained not only until completion of the project but also throughout its transition to operation.

This would include requiring credentials for all personnel on-site as well as security control at all project-access points, consistent with recognized security practices. The resources of the State Police force assigned to the MTA should be considered for this task. More generally, the MTA should coordinate its efforts with those of the Department of Homeland Security for heightening the security of our nation's infrastructure.

7

Conclusions

Pursuant to the committee's statement of task, its review of the project concentrated on five areas: cost and schedule procedures, contract administration, reporting and controls, oversight of the management consultant, and the transition from construction to operations. In all cases the committee's emphasis was on the present situation and the actions necessary to bring the CA/T project to a successful conclusion.

Specific findings and recommendations in these areas are presented in the body of this report. But in general, the committee believes that the current management structure, with the modifications recommended in this report, can successfully complete the project.

The overarching importance of two matters warrants special emphasis. *First,* the project currently has approximately 3500 outstanding claims (about 3200 construction claims and 350 design claims), with an average age of about 600 days. They constitute a large and uncertain financial burden and a potential cause for further delay in closing out the project. These outstanding issues also indicate an inability to reach closure and make contractor payments.

The MTA is aware of the problem, but it needs to move with much more vigor and urgency to resolve it. The MTA should assign adequate staff to this issue and accelerate decision making. The committee believes the long-term savings created by timely resolution of claims will outweigh the short-term expense of possibly adding additional staff.

Resolving the claims is further impeded by the present requirement that claims over $250,000 be automatically sent to the legal staff before the engineers and other technical staff familiar with the details can resolve the issue. This inevitably introduces an adversarial element into the process before the technical issues can be addressed.

Second, the project is entering a phase during which operational elements will be turned over to the operations and maintenance organization. As the committee was drafting this report, the completion date for the I-90 portion of the project (connecting the present turnpike extension to the Ted Williams Tunnel) slipped from its original date in November 2002 and finally accomplished over the weekend of January 18, 2003. The delay was ascribed to problems with instrumentation in the roadway tunnels, including some difficulty obtaining access to appropriate areas. In any case, it illustrated the need for proper organization and management of the transition from construction to operation and maintenance.

The CA/T project is described as the first of many large-scale infrastructure projects that will be needed in the near future; the FHWA alone reports seven current environmental studies for projects that will each be worth over $1 billion. These projects could benefit from the lessons learned from the Big Dig—the causes of the many problems that increased costs and delayed construction, as well as the solutions developed by the management team, design engineers, and construction contractors. Participants in these new projects will need to learn how to develop realistic expectations and manage efforts to achieve them. A detailed, independent, and comprehensive study of the history of the CA/T project and its management would be a prudent investment for the future.

Appendix A

Committee to Review Project Management Practices Employed on the Boston Central Artery ("Big Dig") Project

John T.Christian*(Chair)* is one of the nation's leading geotechnical engineers and an expert in consulting engineering. Dr. Christian spent much of his career at the Massachusetts Institute of Technology and at Stone & Webster Engineering Corporation. He received his bachelor's, master's and doctoral degrees in civil engineering at MIT and later served on its faculty of civil engineering. Dr. Christian was a Vice President at Stone & Webster when he left to go into private practice. He has published over 90 papers and three books in the geotechnical and earthquake engineering fields. He is currently a consulting engineer in Boston and Newton, Massachusetts. Dr. Christian has actively served as a fellow and former chair of the Engineering Accreditation Commission (EAC) of the Accreditation Board for Engineering Technology (ABET), the organization that oversees the accreditation of engineering programs at universities. He is also a former chair of the American Society of Civil Engineers (ASCE) Geotechnical Engineering Division and edited the Society's Journal of Geotechnical and Geoenvironmental Engineering. An Honorary Member of ASCE and an Honorary Member of the Boston Society of Civil Engineers Section, ASCE, Dr. Christian has been the distinguished recipient of several honors and awards. In 1996, he received ASCE's Thomas A. Middlebrook Award for a paper on the uses of reliability approaches in which he applied probabilistic concepts to geotechnical engineering. In 1999 he was elected to the National Academy of Engineering.

Jamie Browder is a civil engineer with over 30 years of experience in virtually all aspects of planning, design, and construction of major civil-works projects, including highways, bridges, commuter rail stations and other transportation elements. Mr. Browder is regional vice president of Wilbur Smith Associates and oversees operations that include civil/roadway design, structural design, construction engineering and inspection, bridge inspection, municipal engineering, traffic engineering, transportation planning, and management of overall staff to ensure conformance to client requirements. Mr. Browder is former Chief Engineer of the Virginia Department of Transportation and throughout his career at VDOT he has served as District Administrator, Assistant District Engineer, and Resident Engineer. His major projects included I-295 and I-95 near Richmond, various roadways and traffic systems within the City of Richmond, and the Woodrow Wilson Bridge Project. Mr. Browder has a B.S. from the Virginia Military Institute.

John C.Davis retired in 2001 from General Motors Corporation's Worldwide Facilities Group, where he was Director of Engineering Services, Capital Projects. Mr. Davis spent 36 years at GM, beginning in the Electrical Department of the Argonaut Division. He worked to become a supervisor and then department head of the Electrical and Mechanical Systems Department. He then became manager of the Project and Construction Administration. Mr. Davis managed the engineering of the Spring Hill, Tennessee, Saturn complex from 1985 to 1989. He then became manager of all of GM's major domestic projects, and later became director of International Facilities, Capital Projects. Mr. Davis is a member of the Industrial Advisory Committee of the

Department of Mechanical and Aerospace Engineering at the University of Missouri, where he graduated with a B.S. in mechanical engineering.

Chris T.Hendrickson is the Duquesne Light Company Professor of Engineering and Head of the Department of Civil and Environmental Engineering at Carnegie Mellon University. His research, teaching, and consulting are in the general area of engineering planning and management, including design for the environment, system performance, project management, finance, and computer applications. He has co-authored two textbooks, *Project Management for Construction* and *Transportation Investment and Pricing Principles,* and two monographs, *Knowledge-Based Process Planning for Construction and Manufacturing* and *Concurrent Computer-Integrated Building Design.* In addition, he has published numerous articles in the professional literature. Dr. Hendrickson is a member of numerous professional organizations and advisory and review boards, and he is editor of the *Journal of Transportation Engineering.* His education includes B.S. and M.S. degrees from Stanford University, a Master of Philosophy degree in economics from Oxford University, and a Ph.D. from the Massachusetts Institute of Technology. Dr. Hendrickson has been the recipient of the ASCE Turner Lecture Award (2002), Fenves Systems Research Award (2002), AT&T Industrial Ecology Fellowships (2000–2002), the ASCE Frank M. Masters Transportation Engineering Award (1994), the ASCE Walter L. Huber Civil Engineering Research Award (1989), the Benjamin Richard Teare Teaching Award from the Carnegie Institute of Technology (1987), and a Rhodes Scholarship (1973).

James Lefter has over 40 years' experience directing design and construction of major facilities in industry and government. He spent 30 years as a structural engineer—in the private sector with Bethlehem Steel, and in the public sector with the General Services Administration and Veterans Administration (VA). He has also served as Project Director/Contracting Officer for Design and Construction for the VA, and as its Director of Architecture and Engineering. While working for the VA, Mr. Lefter created a national seismic-safety program for its facilities and helped develop the first modern seismic design code. He then went on to become the Director of the Learning From Earthquakes Program at the Earthquake Engineering Research Institute. Mr. Lefter has recently been Visiting/Adjunct Professor Lecturer both at the University of Illinois and Virginia Tech, teaching in the area of construction engineering and management. He is a registered professional engineer in Pennsylvania and Maryland and a member of numerous technical and professional societies. Mr. Lefter holds a B.S. and M.S. in civil engineering from the University of Maryland and an M.A. in government from George Washington University.

Richard K.Sandaas is a consultant in project management and former Executive Director of the Municipality of Metropolitan Seattle (METRO). Mr. Sandaas has 40 years of public- and private-sector project experience and is a recognized leader in public-works project management and implementation of major capital programs. He has reviewed major public programs, including Boston's MWRA Deer Island Program; the Clean Waterways Program of Sydney, Australia; the CSO Program of Portland, Oregon; and the Regional Wastewater Program of King County, Washington. Mr. Sandaas has most recently served as Director of the Kingdome Sports Facility, where he was appointed to lead the project to repair and reopen the Kingdome after emergency closure. He has also served as Director of the Technical Services Department, Director of the Transit Department, and Executive Director of the METRO, the largest transit and water-pollution-control agency in the Pacific Northwest. In the private sector, Mr. Sandaas has held positions as project manager for Procter & Gamble and the Boeing Company in plant-facility development. He is a member of the American Public Works Association, which cited him as a "Top Ten Public Works Official," and of the American Public Transit Association. Mr. Sandaas has a B.S. in mechanical engineering from the University of Washington, undergraduate-course experience at MIT, and graduate course work in the University of Washington Management Program.

Richard P.Weaver retired in 1997 as Deputy Director and Chief Engineer for Caltrans (the California Department of Transportation). He held this position for 7 years, having previously worked for the Department for 37 years. During his career there he served as Interim Chief Deputy Director, District Director of District 6 (Fresno), Deputy District Director, and Deputy Project Director of the Sacramento Light Rail Transit Project, among other positions. Presently, Mr. Weaver is a Project Manager at Skillings and Connolly, Inc., for which he manages an office in Ronan, Montana. Mr. Weaver has been a member of various committees at Caltrans, AASHTO (American Association of State Highway and Transportation Officials), and the Transportation Research Board. He chaired AASHTO'S Subcommittee on Traffic Engineering and led its delegation on the rewrite committee of the federal *Manual on Uniform Traffic Control Devices.* He is a registered civil engineer and traffic engineer in California and a registered civil engineer in Montana. Mr. Weaver has a B.S. in civil engineering from Sacramento State University and an M.A, in public administration from San Diego State University. He has a Life Teaching Credential for California Community Colleges and has done graduate coursework in organization management at the University of California, Davis.

Appendix B

Statement of Task

The committee was asked to review and comment on the following issues:

1. The adequacy of the current procedures to balance cost and schedule issues and the degree to which the multiplicity of projects and priorities embodied in the CA/T project are and can be successfully organized and implemented.
2. Process refinements to the organization and management of the contract administration function, including the control of change orders, the contractor claims process, the execution of contract modifications, and the review of cost recovery issues.
3. Recommended improvements to the reporting and controls processes, including determining whether the appropriate amount of information is being provided to project management for decision-making purposes.
4. Adequacy of methods employed by the owner to oversee the management consultant.
5. The appropriateness and effectiveness of the structure that is transitioning the CA/T project from construction to an operable transportation infrastructure system.

Appendix C

Presentations to the Committee

October 21–23,2002 meeting

Project Overview—M.Lewis (Project Director)

Project History- W.Edwards (Program/Budget Manager)

Current Status of the Project—J.Saroufim (Construction Controls Manager)

Project Organization—M.Lewis

Project Development—A.Lancellotti (Deputy Program Manager/Engineering Manager), J. Saroufim

Scope Control—J.Saroufim

Schedule Control—J.Saroufim, J.Ekiert (Construction Schedule Manager)

Cost Control—J.Saroufim, G.Helmich (Construction Cost Manager)

Management Recommendations/Approval Process—M.Lewis, M.Wiley (Program Manager), C.Gottschall (FHWA)

Case Study- J.Allegro (Director of Construction)

November 11–13, 2002 meeting

Claims and Changes Procedure

Original Procedure/Outcome—J.Allegro

Refined Procedure/Outcome—J.Gorman (Director of Claims and Changes)

Budgeting for Claims and Changes (C&C)—J.Saroufim

Owner Oversight and the IPO—M.Lewis, J.Lammie

Governance Roles and Public Accountability—M.Hayman (Deputy Project Director), S.Gee (FHWA), A.Almeida (FHWA)

Monitoring the Management Consultant Work Program—M.Foley (Director of Finance), R. Grenier (Director of Internal Controls)

Monitoring Performance of the Management Consultant—M.Lewis

Monitoring Quality, Safety & Health and Environmental Commitments—J.Allegro, P.Stakutis (Environmental Manager)

Cost Recovery Philosophy—M.Lewis

Cost Recovery Program—K.Dettman (Chief Counsel), M.Breen (Associate General Counsel)

Cost Containment Case Study—M.Lewis

External Oversight—M.Hayman

Systems Commissioning, Turnover and Transition Process—M.Swanson (Chief of Operations/Chief Engineer/Director of Transition), R.Hubbard (Operations Manager), J. Cashman

Value Engineering Change Potential—Gorchev Consultants
December 16,2002 meeting
Roundtable discussion issues:

1) Fairness and adequacy of claims management
2) Effectiveness of partnering
3) Effectiveness of the dispute resolution process
4) Quality of the bid packages.
5) Timeliness and adequacy of responses to requests for information
6) Contract coordination and contractor interface
7) Change management in response to site conditions and design changes
8) Opening schedule and quality control.

Participants:
From Construction Industries of Massachusetts:
-Mike Waters, Slattery
-John Testa, Kiewit
-Bruce Grimaldi, JF White
-Steve Barlow, JF White
-Geoff Collins, Interbeton
-John Pastore, Modern Continental
-Dale Pyatt, Jay Cashman
-John Pourbaix, CIM
From American Council of Engineering Companies of Massachusetts (ACEC/MA):
-Frank Leathers, GEI Consultants (ACEC/MA President)
-David Hatem, Donovan Hatem
-Dom D'Eramo, Rizzo Associates
-Bob Caton, Fay Spofford & Thorndike
-Leo Marino, HNTB
-Skip Skoglund, GPI
From Massachusetts Turnpike Authority:
-Michael Lewis, Project Director
-Joe Allegro, Director of Construction
-Mike Swanson, Chief of Operations/Chief Engineer/Director of Transition
-John Gorman, Director of Claims and Changes
-Kurt Dettman, Chief Counsel
From B/PB (Bechtel and Parsons Brinckerhoff Quade and Douglas):
-Matt Wiley, Program Manager
-Anthony Lancellotti, Deputy Program Manager/Engineering Manager
-John Saroufim, Construction Controls Manager
From Federal Highway Administration:

-Carl Gottschall
From National Research Council:
-Richard Little, Director, Board on Infrastructure and the Constructed Environment (BICE)
-Michael Cohn, Program Officer, BICE
-Jason Dreisbach, Research Associate, BICE
Others:
-Phil Helmes, Intecap

Appendix D

Documents Submitted to the Committee by MTA and IPO

"Summary of EIS (1985)," FHWA-MA-EIS-82–02-F, August 1985.
"Supplemental EIS (1991)," FHWA-MA-EIS-82–02-FS2, January 1991.
"Operations and Maintenance Transfer Documentation—Ramp A—Kneeland (Ramp E-K)," July 2002.
"O&M Transfer Process and Status," received November 2002.
"Deloitte and Touche Report from September 2002," September 2002.
"Organization Chart," November 2002.
"Categories of Changes/Reasons for Changes," November 2002.
"Lemley Report," Peterson Consulting, September 1995.
"Central Artery/Tunnel Project Commitments," December 1997.
"Key Indices Briefing Book," September 2002.
"Roadway Opening Report," November 2002.
"Facilities Opening Report," November 2002.
"Fast Track Study," December 2002.
Detailed listing of 30 active (as of October 31, 2002) construction contracts plus top 10 percent cost-wise.
Listing of second-lowest bidders on construction contracts.
"Boilerplate Construction Contract Document," February 1996.
"Documentation on the Creation of Integrated Project Team," July 1998, July 1999.
"Generic Consultant Contract RFQ/P Boilerplate," August 2002.
"CA/T Contract C19BA Bid Document Package as advertised"

Package of press clippings beginning April 1, 2000.

"Central Artery/Tunnel: Summary of 34 Recommendations," prepared by FHWA, as requested by HR Report 106–622, August 2002.

"Central Artery Claims and Change Process," prepared by Construction Industries of Massachusetts (CIM), December 16, 2002.

"Comments on Improvements for CA/T Project Management," letter to John Christian from American Council of Engineering Companies of Massachusetts (ACEC/MA), December 13, 2002.

"Task Team Review: Management Consultant Staffing," FHWA, September 1994.

"Task Team Review: Time Extensions/Liquidated Damages," FHWA, March 1995.

"Process Review of the Changes and Claims Process," FHWA, September 1998.

"Construction Contract Administration Manual," updated August 2002.

"Central Artery/Tunnel Project Procedures," August 2002.

Examples of Construction Contract Modifications for Contract C09A4.

"Project Engineer's Manual," Revision 4.

"Design Policy Memorandum List and Cross References," October 2002.

"Partnering Manual," January 1998.

Appendix E

BOARD ON INFRASTRUCTURE AND THE CONSTRUCTED ENVIRONMENT

PAUL GILBERT, *Chair,* Parsons Brinckerhoff Quade & Douglas, Seattle, Washington
MASSOUD AMIN, University of Minnesota, Minneapolis
RACHEL DAVIDSON, Cornell University, Ithaca, New York
REGINALD DESROCHES, Georgia Institute of Technology, Atlanta
DENNIS DUNNE, California Department of General Services, Sacramento
PAUL FISETTE, University of Massachusetts, Amherst
YACOV HAIMES, University of Virginia, Charlottesville
HENRY HATCH, U.S. Army Corps of Engineers (retired), Oakton, Virginia
AMY HELLING, Georgia State University, Atlanta
SUE MCNEIL, University of Illinois, Chicago
DEREK PARKER, Anshen+Allen, San Francisco, California
DOUGLAS SARNO, The Perspectives Group, Inc., Alexandria, Virginia
WILL SECRE, Masterbuilders, Inc., Cleveland, Ohio
DAVID SKIVEN, General Motors Corporation, Detroit, Michigan
MICHAEL STEGMAN, University of North Carolina, Chapel Hill
DEAN STEPHAN, Charles Pankow Builders (retired), Laguna Beach, California
ZOFIA ZAGER, County of Fairfax, Fairfax, Virginia
CRAIG ZIMRING, Georgia Institute of Technology, Atlanta

Staff

RICHARD G.LITTLE, Director, Board on Infrastructure and the Constructed Environment
LYNDA L.STANLEY, Executive Director, Federal Facilities Council
MICHAEL COHN, Program Officer
DANA CAINES, Financial Associate
JASON DREISBACH, Research Associate
PAT WILLIAMS, Senior Project Assistant